Date Due

FEB 1 7 2003	JUL 7 2009	
APR 1 2003	DEC 1 4 2009	
MAY 5 2003	FEB 0 3 2010	
SEP 2 2003	MAR 9 2010	
OCT 3 1 2003	APR 1 4 2010	
NOV 2 4 2003	AUG 2 3 2010	
MAR 5 2004	MAR 1 9 2011	
AUG 2 0 2004	AUG 3 0 2011	
SEP 5 2007	JAN 1 8 2012	
OCT 1 9 2007	MAY 9 2012	
MAR 1 4 2008	SEP 3 0 2013	
MAY 2 3 2008	NOV 1 8 2015	

UNDER ONE ROCK

Bugs, Slugs and other Ughs

By Anthony D. Fredericks

Illustrated by Jennifer DiRubbio

Dawn Publications

Dedication

For my wife, Phyllis, and my children,
Rebecca and Jonathan–they are the sunshine of my life. – A.D.F.

To my son, Zachary, for playing so nicely while I worked.
Thank you to my husband, Rob, my family and friends for all your help.
I couldn't have done it without you. - JDR

    Library of Congress Cataloging-in-Publication Data

Fredericks, Anthony D.
        Under one rock : bugs, slugs, and other ughs / by Anthony D.
Fredericks ; illustrated by Jennifer DiRubbio.
        p. cm. — (A sharing nature with children book)
        ISBN 1-58469-027-5 (pbk.) — ISBN 1-58469-028-3 (hardback)
        1. Soil animals—Juvenile literature. [1. Soil animals.]   I.
DiRubbio, Jennifer, ill. II. Title. III. Series.
QL110 .F74 2001
591.75'7—dc21
        2001002038

Dawn Publications
P.O. Box 2010
Nevada City, CA 95959
530-478-0111
nature@dawnpub.com

Printed in China

10 9 8 7 6 5 4 3 2 1
First Edition
Design and computer production by Andrea Miles

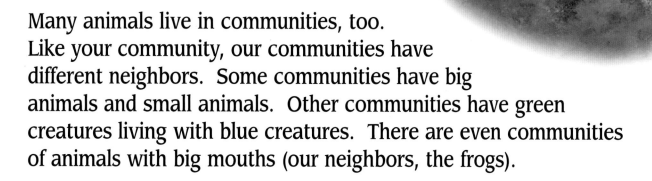

Dear Neighbors,

There are many places to live.
You may live in a large city.
Or, you may live in a small town.
You might live on a busy street or
a quiet road. Wherever you live,
you are part of a community.

Many animals live in communities, too.
Like your community, our communities have
different neighbors. Some communities have big
animals and small animals. Other communities have green
creatures living with blue creatures. There are even communities
of animals with big mouths (our neighbors, the frogs).

We animals have learned something important. We may be
different, but we all live in the same space. This space is called a
habitat. A habitat is the area where certain animals or plants
normally live. It means we have links that may help us survive.
It also means that what affects one of us may affect all of us.

I hope you enjoy visiting our special community. It's a place with
many neighbors. To learn about one of us is to know about all of
us. That's because we all live together.

Your eight-eyed friend,

Spider

Here is a field for insects to play in
  And acres of shade for turtles to lay in;
With wind-brushed trees for birds to nest in,
  And sun-splashed spaces for lizards to rest in.
This is where, on this summer day,
  There lay a rock, all rough and gray.

This is the rock.

The rough-gray rock was discovered by chance
By a brown-skinned boy in ragged pants—
A curious lad who wondered aloud,
"What could be hiding in the red-rich ground?"

He lifted the rock, all rough and gray
    That he saw in the field on that summer's day,
And there he found some varied creatures,
    A village of animals with special features.

These are the **earthworms** all squiggly and round
Who aerate the soil in the red-rich ground,
Below the big rock, all rough and gray
That hides a whole crowd on a summer's day.

This is the army of hundreds of **ants**
Who dig twisting tunnels and farm tiny plants,

Neighbors to **earthworms** all squiggly and round
Who aerate the soil in the red-rich ground,
Below the big rock, all rough and gray
That hides a whole crowd on a summer's day.

This is the **spider** with her eight-eyed face
Who builds a home in this cool dark place,
A home near the army of hundreds of **ants**
Who dig twisting tunnels and farm tiny plants,
Neighbors to **earthworms** all squiggly and round
Who aerate the soil in the red-rich ground,
Below the big rock, all rough and gray
That hides a whole crowd on a summer's day.

This is the **beetle** all shiny and black
  With grooves running down both sides of his back.
A friend of the **spider** with her eight-eyed face,
  They live side by side in this cool dark place,
A home near the army of hundreds of **ants**
  Who dig twisting tunnels and farm tiny plants,
Neighbors to **earthworms** all squiggly and round
  Who aerate the soil in the red-rich ground,
Below the big rock, all rough and gray
  That hides a whole crowd on a summer's day.

Some tiny **field crickets** who sing with their feet
Search near the rock for some seeds they can eat.
They live with the **beetle** all shiny and black
With grooves running down both sides of his back.

A friend of the **spider** with an eight-eyed face
   They live side by side in this cool dark place,
A home near the army of hundreds of **ants**
   Who dig twisting tunnels and farm tiny plants,
Neighbors to **earthworms** all squiggly and round
   Who aerate the soil in the red-rich ground,
Below the big rock, all rough and gray
   That hides a whole crowd on a summer's day.

A sole **millipede** with a sensitive feel
  Slips through the dirt in search of a meal.
He plows by the **crickets** who sing with their feet,
  And search near the rock for some seeds they can e
They live with the **beetle** all shiny and black
  With grooves running down both sides of his back.

He's a friend of the **spider** with her eight-eyed face,
  They live side by side in this cool dark place,
A home near the army of hundreds of **ants**
  Who dig twisting tunnels and farm tiny plants,
Neighbors to **earthworms** all squiggly and round
  Who aerate the soil in the red-rich ground,
Below the big rock, all rough and gray
  That hides a whole crowd on a summer's day.

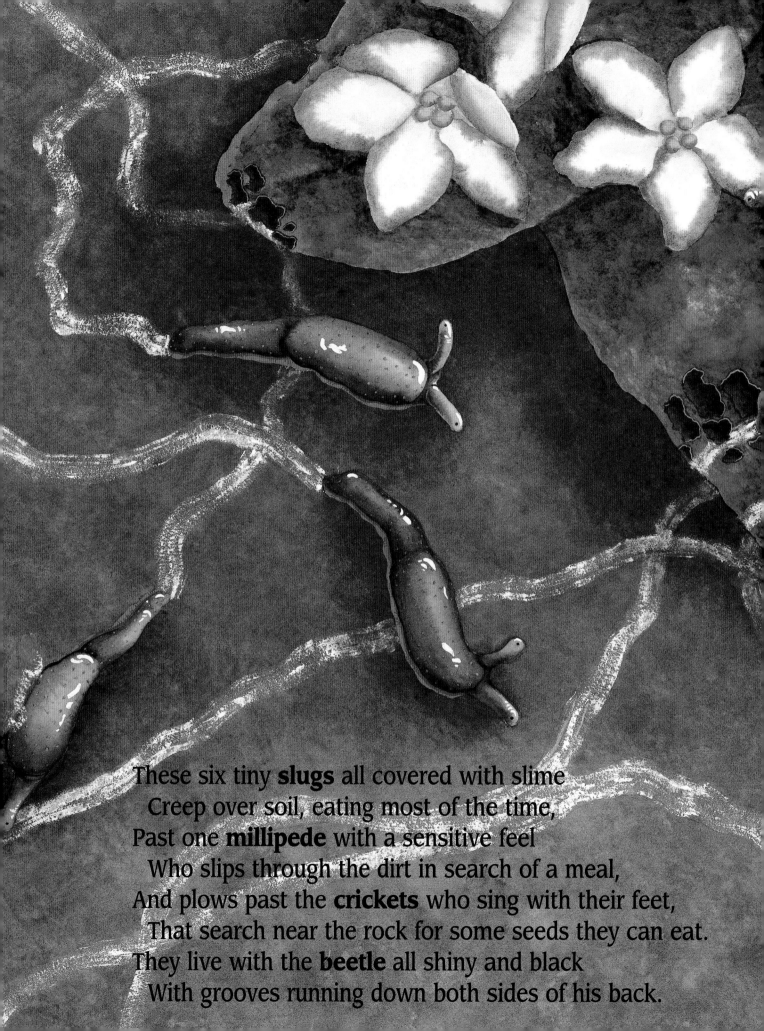

These six tiny **slugs** all covered with slime
Creep over soil, eating most of the time,
Past one **millipede** with a sensitive feel
Who slips through the dirt in search of a meal,
And plows past the **crickets** who sing with their feet,
That search near the rock for some seeds they can eat.
They live with the **beetle** all shiny and black
With grooves running down both sides of his back.

He's a friend of the **spider** with her eight-eyed face,
    They live side by side in this cool dark place,
A home near the army of hundreds of **ants**
    Who dig twisting tunnels and farm tiny plants,
Neighbors to **earthworms** all squiggly and round
    Who aerate the soil in the red-rich ground,
Below the big rock, all rough and gray
    That hides a whole crowd on a summer's day.

The creatures and critters live together as one,
Beneath the gray rock, away from the sun.
A collection of neighbors – the large and the small;
And the place where they live is home to them all.

Field Notes

Spiders, slugs, ants and the other creatures in this book can be found throughout the world. The specific species described and illustrated in this book are all native to North America. Their habitat, like other animal habitats, offers a fascinating look into how various ecosystems, food chains or food webs work.

Earthworms are found all over the world in all types of soil. Most move through the ground eating dead fragments of plants along with the soil. The soil passes through their bodies and is deposited in new locations. This not only aerates the soil, but also brings new dirt to the surface. Earthworms are important to farmers and to the life cycle of many plants.

Fantastic Fact: The largest species of earthworms live in Australia. They often grow to lengths of nine feet or more.

Ants have lived on the Earth for more than 100 million years. They are some of the most widespread of all animals. Ants typically live in colonies that are divided into groups. These groups may include "soldiers" (ants who guard and protect the colony), "nurses" (ants who take care of the newborn members of the colony), and "farmers" (ants who tend enormous underground gardens). Many species of ants eat small mushrooms and fungi cultivated deep within their tunnels.

Fantastic Fact: Ants can lift 50 times their own body weight—with their mouths.

There are more than 35,000 species of spiders living throughout the world. Spiders have four pairs of legs; whereas insects have three pairs of legs. Most spiders have eight simple eyes, although some species have fewer. Interestingly, spiders have bad eyesight. Some spiders have such poor vision that they cannot find an insect that is right in front of them. If the insect moves, however, the spider can detect the vibrations it makes and pounce on it. Spiders can be found in trees and bushes, along the ground, and even underground.

Fantastic Fact: The largest spider in the world is the bird-eating spider of South America. It grows to the size of a dinner plate.

Beetles

There are about 300,000 species of beetles worldwide. Some crawl on land, others fly, and a few live in water. Their tough, armor-like wings distinguish them from other insects. Some species feed on plants. Others eat various plant pests. Beetles have chewing mouthparts that enable them to eat many things. Some species of beetles live entirely on dead animals.

Fantastic Fact: When threatened, the bombardier beetle shoots a spray from its rear end—at up to 500 times a second.

Field Crickets

Field crickets are black to dark reddish brown. They grow to lengths of approximately five-eighths to one inch. Crickets like to live in areas where there is protection from night winds and cold. Their diet includes seeds, small fruits, and dead insects. They are noted for their songs—usually a series of triple chirps. Male crickets produce this sound by rubbing together roughened portions of their wings or legs. These sounds are used in courtship, to establish territory, or as a form of warning.

Fantastic Fact: During courtship, male field crickets dance around and "sing" to excite the females.

Millipede

The word millipede means thousand-footed. However, no millipede has 1,000 feet. Some have as few as 20 legs, while a few tropical species have as many as 230 legs. Their legs are designed for moving through loose soil and humus. Millipedes have two pairs of legs on each body segment. Centipedes, on the other hand, have one pair of legs per body segment. Millipedes have two very sensitive antennae on their heads. These antennae have sensors that can taste foods, smell odors, measure temperature, and find water. Most millipedes eat dead plants and wood.

Fantastic Fact: One species of millipede grows up to 11 inches long. When it is disturbed it coils up into a ball the size of a golf ball.

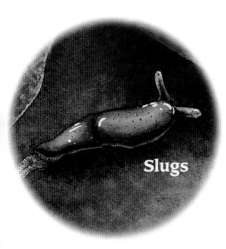

Slugs

Slugs crawl on their bellies, so they are known as gastropods, a word which means "belly foot." Some slugs live in the ocean, some live on land. Land slugs are covered with gooey slime. This slime helps them glide over rough surfaces and sharp objects. Slugs spend most of the day underground or under rocks. This prevents them from drying out. Many species of slugs are herbivorous. That means they eat mostly stems, leaves, and roots of plants. When baby slugs are born they have to take care of themselves. They must find their own food and protect themselves from enemies.

Fantastic Fact: Some species of slugs have more than 20,000 teeth.

How to Learn More

Dear Reader,

Ecology is the study of animals and their environment. Here are some of my favorite resources about this fascinating field of science.

Anthology for the Earth edited by Judy Allen, a wonderful collection of art, essays and poetry about the planet earth.

This is the Sea that Feeds Us by Robert Baldwin, a colorful book about special creatures in the ocean.

The Best Book of Bugs by Claire Llewellyn, a book filled with loads of incredible information about some of the most amazing creatures in the animal kingdom.

Night Letters by Palmyra LoMonaco, a wonderful book about a young girl who takes notes on what the backyard insects and natural objects have to tell her.

Bugs by Nancy Winslow Parker and Joan Richards Wright, a delightful collection of 16 insects in funny rhymes, informative drawings, and solid facts.

The Tree in the Ancient Forest by Carol Reed-Jones, a delightful book about the web of plants and animals in an old tree.

Millipedeology by Michael Elsohn Ross, an up-close-and-personal introduction to some extraordinary creatures.

Look!: The Ultimate Spot-the-Difference Book by April Wilson, a wordless picture book about some of the Earth's diverse natural habitats.

Here are some of the other children's books I've written.

Animal Sharpshooters, animals that throw lassos (bolas spider), squirt blood from their eyes (horned toads), or spit at their dinner (archer fish).

Cannibal Animals, an incredible collection of some of the most unusual and amazing animals—those that eat their own kind.

Clever Camouflagers, an amazing book about how some clever ways animals hide from their predators.

Slugs, a book that offers some incredible information and eye-popping photographs about these wonderful creatures.

Surprising Swimmers, a book about squids, sea serpents, and birds that fly underwater.

Weird Walkers, a book about a lizard that walks on water, a fish that walks on land, and an animal that walks upside down.

Here are the names and addresses of organizations working hard to preserve animal habitats. You might want to contact them to find out what they are doing and how you can become involved.

Friends of Wildlife Conservation
New York Zoological Society
185 Street, Southern Blvd.
Bronx, NY 10460
www.wcs.org

National Audubon Society
700 Broadway
New York, NY 10003
www.audubon.org

National Wildlife Federation
11100 Wildlife Center Drive
Reston, VA 20190
www.nwf.org

Nature Conservancy
1815 North Lynn Street
Arlington, VA 22209
www.nature.org

If you or your teacher would like to learn more about me and the books I write, please log on to my web site, www.afredericks.com

Anthony D. Fredericks is a nature explorer from way back. He beachcombed and poked around in tide pools near his native Newport Beach, California, and during summers explored the Sierra Nevada mountains of eastern California. Later he explored the high deserts of Arizona while attending school there. Now Anthony explores the mountainside in Pennsylvania where he and his wife live. A former classroom teacher and reading specialist, he is Professor of Education at York College. As the author of more than 20 children's books he is a frequent visitor to schools around the country, where he shares the wonders of nature with a new generation of naturalists.

Jennifer DiRubbio is both a passionate artist and an avid environmentalist. She has been active as an artist for several organizations that promote nature and a healthy planet. Jennifer graduated with a BFA from Pratt Institute in 1992. She keeps her home and studio in Merrick, New York, as "green" and environmentally sound as possible, where her husband and young child also work and play.

A SAMPLING OF NATURE AWARENESS BOOKS FROM DAWN PUBLICATIONS

Birds in Your Backyard by Barbara Herkert, is an excellent tool to help kindle the spark of interes in birds at an early age, and so establish a life-long passion for nature in general, and birds in particular.

A Tree in the Ancient Forest by Carol Reed-Jones. The plants and animals around a grand old fir are remarkably and wonderfully dependent upon each other. Christopher's magical realism, and Carol's cumulative verse serve both to inform the mind and inspire the soul.

This is the Sea that Feeds Us by Robert F. Baldwin. In simple cumulative verse, beginning with tiny plankton, "floating free," this book explores the oceans' fabulous food chain that reaches al the way to whales and humans in an intricate web.

Stickeen: John Muir and the Brave Little Dog by John Muir as retold by Donnell Rubay. In this classic true story, the relationship between the great naturalist and a small dog is changed forever by their adventure on a glacier in Alaska.

Wonderful Nature, Wonderful You by Karin Ireland, shows how nature is a great teacher, reminding us to bloom where we are planted. This popular book celebrates nature's diversity and strong character traits—and with a light touch, suggests how humans can follow nature's example and make good choices.

John Muir: My Life with Nature by Joseph Cornell. John Muir's remarkable adventures and attunement with nature are told in his own words, edited for a young audience by the author of the classic *Sharing Nature with Children*.

Do Animals Have Feelings, Too? by David Rice, presents fascinating true stories of animal behavior, and then asks the reader whether they think the animal is acting on the basis of feelings or just instinctively.

Animal Acrostics by David M. Hummon. Acrostic poems are a wonderful way to encourage children to write creatively. These "vertical poems" are amusing, clever, and informative.

Salamander Rain, A Lake and Pond Journal by Kristin Joy Pratt-Serafini. This young author-illustrator's fourth book is a "planet scout's" lake and pond journal—a subject known to be wet and muddy, but fun! Kristin is the teenage "eco-star" made famous by her books *A Walk in the Rainforest, A Swim through the Sea,* and *A Fly in the Sky.*

Dawn Publications is dedicated to inspiring in children a deeper understanding and appreciation for all life on Earth. To order, or for a copy of our catalog, please call 800-545-7475. Please also visit our web site at www.dawnpub.com.